SpringerBriefs in Molecular Science

Green Chemistry for Sustainability

Series Editor
Sanjay K. Sharma

For further volumes:
http://www.springer.com/series/10045

Mika Sillanpää · Thuy-Duong Pham
Reena Amatya Shrestha

Ultrasound Technology
in Green Chemistry

 Springer

Mika Sillanpää
Laboratory of Green Chemistry
LUT Faculty of Technology
Lappeenranta University of Technology
Patteristonkatu 1
50100 Mikkeli
Finland
e-mail: Mika.Sillanpaa@lut.fi

Reena Amatya Shrestha
Department of Civil and Environmental
  Engineering
Lehigh University
13 E. Packer Ave
Bethlehem 18015
USA
e-mail: reenashrestha@yahoo.com

Thuy-Duong Pham
Laboratory of Green Chemistry
LUT Faculty of Technology
Lappeenranta University of Technology
Patteristonkatu 1
50100 Mikkeli
Finland
e-mail: duong.pham@uku.fi

ISSN 2191-5407
ISBN 978-94-007-2408-2
DOI 10.1007/978-94-007-2409-9
Springer Dordrecht Heidelberg London New York

e-ISSN 2191-5415
e-ISBN 978-94-007-2409-9

Cover design: eStudio Calamar, Berlin/Figueres

Printed on acid-free paper

Springer is part of Springer Science+Business Media (www.springer.com)

Contents

Ultrasound Technology in Green Chemistry

Abstract As part of a new and rapidly growing field of study, the applications of ultrasound in green chemistry and environmental applications have a promising future. Compared to conventional methods, ultrasonication can bring various benefits, such as environmental friendliness (no toxic chemicals are used or produced), cost efficiency, and compact, on-site treatment. Besides an overview of the ultrasonic background, this paper summarizes the main findings and innovations of recent studies that have been using ultrasound in environmental analysis, water and sludge treatment, and soil and sediment remediation, as well as air purification.

Keywords Ultrasound · Environmental analysis · Water treatment · Sludge stabilization · Soil remediation · Air purification · Green chemistry

1 Introduction

Ultrasound refers to inaudible sound waves with frequencies in the range of 16 KHz–500 MHz, above the upper limit of human hearing. It can be transmitted through any elastic medium including water, gas-saturated water, and slurry. Ultrasound has been used for diverse purposes in many different areas (Fig. 1).

In terms of frequency, ultrasound can be categorized into two main strands: (1) high frequency (2–10 MHz)—low power diagnostic ultrasound, involving medical imaging, non-destructive testing, and (2) low to medium frequency (20–1000 kHz) frequency—high power ultrasound, involving other applications in industry, nanotechnology, ultrasonic therapy and sonochemistry. Among these various utilizations, this paper will focus on the uses of ultrasound in the main areas of environmental science and technology relevant to green chemistry, from water and sludge treatment, soil and sediment remediation, and air purification to environmental analysis (Fig. 2).

M. Sillanpää et al., *Ultrasound Technology in Green Chemistry*,
SpringerBriefs in Green Chemistry for Sustainability,
DOI: 10.1007/978-94-007-2409-9_1, © The Author(s) 2011

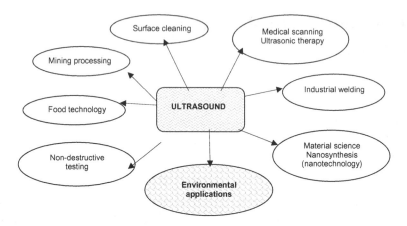

Fig. 1 Diverse applications of ultrasound

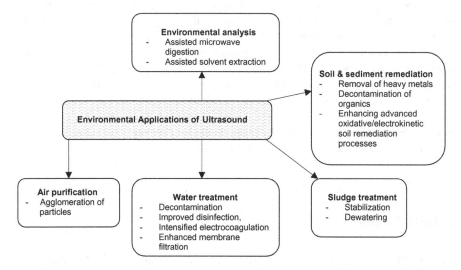

Fig. 2 Ultrasound in environmental applications

Although ultrasonic applications in environmental areas are still in lab-scale and developing stage, they are increasing rapidly and attracting more and more interest because of many advantages they offer: environmental friendliness (no toxic chemicals are used or produced), low energy demand, and compact and transportable method that can be used on-site. Environmental remediation by ultrasonication is involved mostly inorganic pollutant destruction, through thermal decomposition (pyrolysis) and the formation of oxidative species, like hydroxyl radicals, which enhance the mineralization of pollutants. Moreover, in soil treatment, ultrasonic waves increase the porosity of the soil and percolation rate, thus accelerating desorption, and facilitating the removal of entrapped contaminants.

Fig. 3 Three reaction zones in the cavitation process

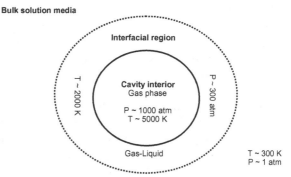

On the other hand, ultrasound applied in environmental analysis also provides benefits such as shorter time, simplified procedure, and higher purity of the final product. In this book, we have updated our earlier work with the most recent developments (Pham et al. 2009a, b).

2 Ultrasound: Background Overview

Ultrasound, like any sound wave, is propagated via a series of compression and rarefaction waves induced in the molecules of the medium through which it passes. Compression cycles push molecules together, while expansion cycles pull them apart. At sufficiently high power, the rarefaction cycle may exceed the attractive forces of the molecules of the liquid, and cavitation bubbles will form. Cavitation bubble collapse is a remarkable phenomenon induced throughout the liquid. Cavitational collapse produces intense local heating (~ 5000 K) and pressures (~ 1000 atm) with very short lifetimes, implying the existence of extremely high heating and cooling rates ($>10^9$ K/s). It has been shown that transient supercritical water is obtained during the collapse of cavitation bubbles generated sonolytically (Hoffmann et al. 1996). Acoustic cavitation provides a unique interaction of energy and matter, and ultrasonic irradiation of liquids causes high energy chemical reactions to occur (Suslick 2006).

According to Adewuyi (2001), so far four theories have been proposed to explain the sonochemical events: "hot-spot" theory, "electrical" theory, "plasma discharge" theory, and "super critical" theory. These have led to several modes of reactivity being proposed: pyrolytic decomposition, hydroxyl radical oxidation, plasma chemistry, and super critical water oxidation. Generally, most studies in environmental sonochemistry have adopted the "hot-spot" concepts to explain experimental results. In the hot-spot model (Adewuyi 2001), three regions are postulated (Fig. 3): (1) a hot gaseous nucleus, (2) an interfacial region, and (3) a bulk solution at ambient temperature. Reactions involving free radicals can occur within the collapsing bubble, at the interface of the bubble, and in the surrounding liquid.

Within the center of the bubble, high temperatures and pressures generated during cavitation provide the activation energy required for bond breakage, dissociation of solvents and other vapors or gases, leading to the formation of free radicals or excited species. The radicals generated either react with each other to form new molecules and radicals or diffuse into the bulk liquid to serve as oxidants.

The second reaction site is the liquid shell immediately surrounding the imploding cavity, which has been estimated to heat up to approximately 2000 K during cavity implosion. In this solvent layer surrounding the hot bubble, both combustion and free-radical reactions (involving ·OH derived from the decomposition of H_2O) occur. Reactions here are comparable to pyrolysis reactions. Pyrolysis in the interfacial region is predominant at high solute concentrations, while at low solute concentrations, free-radical reactions are likely to predominate. It has been shown that the majority of degradation takes place in the bubble-bulk interface region.

In the bulk liquid, no primary sonochemical activity takes place, although subsequent reactions with ultrasonically generated intermediates may occur. A small number of free radicals produced in the cavities or at the interface may move into the bulk-liquid phase and react with the substrate present there in the secondary reactions to form new products. Depending on their physical properties and concentrations, molecules present in the medium will be burned in close to the bubble (pyrolysis) or will undergo radical reactions.

These chemical effects (sonochemistry) explained above are utilized in most of the ultrasonic applications in environmental remediation, especially in organic decontamination. In addition to that, the physical (mechanical) effects of ultrasound are also useful in some environmental applications like air purification, sludge dewatering, and metal leaching. Studies that applied ultrasound in environmental science and engineering with the focus on the most recent ones will be summarized and discussed in more detail in the next sections of this book.

3 Water Treatment

In water treatment technology, the applications of ultrasound (ultrasonication) can be useful in various processes like organic decontamination, disinfection, electrocoagulation, and membrane filtration.

3.1 General Aspects

Due to the cavitation phenomenon, the formation of free radicals and high localized temperatures and pressures, ultrasonic irradiation (ultrasonication) appears to be an effective method for the destruction of hazardous organic compounds in water (Hoffmann et al. 1996; Joseph et al. 2000). The beneficial effect of ultrasonication on the removal of several target compounds from aqueous solutions has been demonstrated in many studies. These compounds include phenol (Enterazi et al. 2003), chlorophenols, nitrophenols, aniline (Emery et al. 2003; Teo et al. 2001; Jiang et al. 2002a, b; Papadaki et al. 2004; Goskonda et al. 2002; Wang et al. 2011), trichloroethylene (Drijvers et al. 1996), perchloroethylene (Saez et al. 2011), ethylbenzene (Visscher et al. 1997), chlorobenzene (Dewulf et al. 2001), cationic surfactant laurylpyridinium chloride (Singla et al. 2011), chloronaphthalene (Jiang et al. 2002a, b), polychlorinated biphenyls, pesticides, polycyclic aromatic hydrocarbons, azobenzene, textile dyes (Joseph et al. 2000, Tezcanli-Guyer and Ince 2003), organophosphate pesticide parathiom (Yao et al. 2010), carbofuran (Hua and Pfalzer-Thompson 2001), nitroaromatics (Abramov et al. 2006), hydrazine (Nakui et al. 2007), diclofenac (Naddeo et al. 2010), polyaromatic hydrocarbons naphthalene, phenanthrene and pyrene (Manariotis et al. 2011), and detergents and surfactants (Adewuyi 2001, Sister and Kirshankova 2005, Abu-Hassan et al. 2006, Belgiorno et al. 2007). Among these various organic contaminants, phenol and phenolic compounds are the most widely investigated. Many studies on sonodegradation of phenolic compounds are summarized carefully in two interesting reviews of Kidak and Ince (2006) and Gogate (2008). In general, the optimum range for the frequency lies between 200 and 540 kHz, while the best pH is in the acid region (Kidak and Ince 2006).

Studying the effect of pH, Jiang et al. (2002a, b) concluded that the pH of a solution plays an important role in the rate of polar aromatic compound degradation by sonolysis, because it affects the charge of the substances (negatively charged under alkaline conditions like 4-nitrophenol, or positively charged at acidic pH like aniline). For these hydrophilic compounds, the neutral species are more easily diffused to and accumulated at the hydrophobic interface of liquid–gas bubbles in comparison with their corresponding ionic forms. Thus, the rate of 4-nitrophenol degradation decreases with increasing pH, while the rate of aniline destruction exhibits a maximum under alkaline conditions. The ultrasonic induced formation of H_2O_2 is also affected by the pH as the yield of H_2O_2 has a maximum at a pH of approximately 3 and decreases with increasing pH (Jiang et al. 2002a, b).

Ultrasonic irradiation of carbofuran ($C_{12}H_{15}NO_3$) was performed at 16 and 20 kHz by Hua and Pfalzer-Thompson (2001) showing that the rate of carbofuran decomposition increased with higher power density applied (1.65–5.55 W/mL), lower initial carbofuran concentrations (25 μM vs. 130 μM), and when sparging with an Ar/O_2 mixture.

Low frequency at 20 kHz has also been used for sonodegradation of linear alkylbenzene sulfonate (LAS) solutions (Abu-Hassan et al. 2006). 20 kHz ultrasound demonstrated a capability of degrading the sodium dodecylbenzene sulfonate (SDBS, a representative LAS molecule). However, the complete mineralisation may not be possible. Degradation rates increased with increasing power and decreasing temperature, and the volume of samples.

The effect of low frequency (20 kHz) ultrasonication on the removal of SDBS and phenolic compounds, with and without several heterogeneous catalysts (Pt, Pd, Ru, CuO.ZnO) was investigated by Papadaki et al. (2004). Such process was called sonocatalytic oxidation. The performance indicated that among these catalysts, a CuO.ZnO supported on alumina catalyst appears to enhance the SDBS decomposition and total oxidation rates, as well as hydrogen peroxide formation. By ultrasonication alone, phenolic compounds at an initial concentration of 0.1 g/l were only partially removed, with about 10–20% removals after 180 min irradiation. However, in the presence of Fe^{2+} ions at concentrations as low as 10^{-3} g/l, the rate of sonolytic degradation increased generally more than 2.5 times (Papadaki et al. 2004).

Studying the frequency effect on the sonochemical remediation of alachlor, a widely employed herbicide, Wayment and Casadonte (2002) concluded that in general, the frequency of 300 kHz provides a faster rate of degradation than either a higher (446 kHz) or lower (20 kHz) frequency under comparable energy input. The study also examined the effects of dissolved gases, and the results indicated that Argon-saturated solutions displayed an enhancement in the degradation rate by a factor of two compared to either oxygen- or air-saturated solutions upon sonication at 300 kHz (Wayment and Casadonte 2002).

Ultrasonic irradiation at approximately 500 kHz has been investigated for the degradation of textile dyestuff by Joseph et al. (2000) and Tezcanli-Guyer and Ince (2003). Both studies presented that oxidation by ultrasonically generated hydroxyl radicals was the main mechanism responsible for dye degradation. In general, sonochemical bleaching is a fast process, as complete decolorization was achieved within 40–150 min, while mineralization is attained after a longer period of time (Joseph et al. 2000). However, total detoxification could be achieved within shorter contact than complete mineralization (Tezcanli-Guyer and Ince 2003). A summary of many studies utilizing ultrasound in the decolouration and mineralization of textile dyes is presented in the review by Vajnhandl and Le Marechal (2005).

The application of high frequency ultrasound at 2.4 and 1.7 MHz was successfully performed to remove ammonia from simulated industrial wastewater, within a short time of 1.5–2 h, by Matouq and Al-Anber (2007).

Entezari et al. (2003) conducted ultrasonication experiments with the same ultrasonic power (50 W) in three different devices operating at 20, 35 and 500 kHz to remove phenol from water. The 35 kHz reactor is a new cylindrical tube called Sonitube™, the results of which were compared to the results of the other two classical devices. The results showed that, without oxidant addition, the rate of phenol destruction was higher at 500 kHz than at 35 or 20 kHz. However, with the addition of hydrogen peroxide, the rate of phenol decomposition was higher at 35 kHz than at 500 or 20 kHz. It was explained that such a different behaviour was not necessarily a pure frequency effect, but could be due to a response to other factors like the surface area of the sonicator (acoustic field) and intensity (Entezari et al. 2003). The coupling oxidant-ultrasound method proved to be more effective than the ultrasound or the oxidant alone. The sonochemical degradation (20 kHz) of phenol and other phenolic pollutants such as chlorophenol and dichlorophenol was also carried out in the work of Emery et al. (2003). Results showed that the relative degradation decreased in the order: 2-chlorophenol > 3,4-dichlorophenol > phenol, and the removal rates increased in the presence of Fe^{2+} ions.

Most of the researches on sonochemistry using continuous ultrasonic irradiation, but Casadonte et al. (2005), though only still an initial study, bring an interesting and new perspective to the utilization of pulse mode ultrasonic irradiation. They explored the application of power-modulated pulsed (PMP) ultrasound in the degradation of acid orange, a common azo-dye used as an industrial colorant. The performances indicated that PMP ultrasound may be more effective in terms of producing hydroxyl radicals and other oxidants in aqueous media, thus, the degradation rate increased by a factor of three compared to continuous irradiation under equal acoustic input power (Casadonte et al. 2005).

Sonication could be used as a pre-treatment step to enhance the biodegradability of the effluent by transforming the molecules into simpler ones, which are further degraded by microorganisms in the following biological treatment step (Sangave and Pandit 2004). Alternatively, sonication could possibly be employed as a post-treatment step for the removal of several refractory compounds commonly formed during thermochemical and advanced oxidation processes (Emery et al. 2003). Ultrasonication can further increase the degradation of intermediates into smaller molecules, even into carbon dioxide and water (Teo et al. 2001). If environmental applications of ultrasonic techniques emerge as post-treatment schemes for the destructive removal of refractory compounds in effluent streams, they will mostly require the use of medium-frequency ultrasound, since such chemicals are usually macromolecules with complex molecular structures and hydrophilic properties. It is fortunate that the reactor systems designed for medium frequency irradiation are relatively easier to maintain than those operated with power ultrasound, due to the drawbacks associated with the latter as noise and cavitational erosion. Such problems, however, may be overcome by sound-proof material and the proper selection, configuration and maintenance of the equipment (Ince et al. 2001).

Most studies indicate that ultrasonication alone cannot be an economical technique for wastewater treatment but it should be combined with or work as an enhancement to some conventional methods. Hybrid methods such as ultrasound/ H_2O_2 or O_3, ultrasonication assisted by catalysts/additives, sonophotocatalytic oxidation, sonoelectrochemistry, and ultrasonication coupled with biological oxidation have been discussed carefully in Gogate's review (2008).

Ozonation combined with ultrasonication in the degradation of p-Aminophenol (PAP) in an aqueous solution was investigated recently by He et al. (2007). This combination of ultrasound and ozone resulted in a synergetic increase in the overall process rate. Although ozonation (72 and 90% PAP removal at 10 and 30 min, respectively) was more effective than ultrasonication (3 and 4% at 10 and 30 min). The efficiency of the combination of ozone and ultrasound (88 and 99% at 10 and 30 min) exceeded even the sum of those using ozone and ultrasound alone. The synergy observed in the combined treatment was explained to be mainly due to the effects of sonolysis in enhancing the decomposition of ozone in collapsing bubbles to yield additional free radicals (He et al. 2007).

Yasman et al. (2004) developed a new method for the detoxification of hydrophilic chloroorganic pollutants (common herbicide 2,4-dichlorophenoxy-acetic acid (2,4-D) and its derivative 2,4-dichlorophenol) in effluent water, using a combination of ultrasound waves, electrochemistry, and Fenton's reagent. The high degradation power of this process is due to the large production of oxidizing hydroxyl radicals and the high mass transfer made by sonication. The application of this sono-electrochemical Fenton process (SEF) treatment (at 20 kHz) with quite a small current density, accomplished almost 50% oxidation of 2,4-D solution (300 ppm, 1.2 mM) in just 60 s. Similar treatments ran for 10 min resulted in a practically full degradation of the herbicide. Thus, the efficiency of the SEF process is much higher than the efficiency of the other methods, and the time required for the full degradation considerably shorter. However, the oxidative degradation of 2,4-D was accompanied by the production of the highly toxic intermediate 2,4-dichlorophenol. Therefore Yasman et al. (2006) reported another study in the enhanced sono-electro-degradation of these chloroorganic compounds by catalysts Pd or Pd/Fe (promoters for reductive chlorine atom abstraction). The bimetallic Pd/Fe catalyst performed superiorly to the pure Pd catalyst. Their findings demonstrate that coupling ultrasound to the electro-catalytical reduction of 2,4-D results in the complete mineralization of the substrate with remarkably shorter reaction times compared to traditional electro-catalytic processes. It was concluded that this method is promising for the remediation of both fresh and wastewaters contaminated by chloroorganic compounds (Yasman et al. 2006).

Gao et al. (2011) developed an efficient Er^{3+}:$YAlO_3$/TiO_2-ZnO composite material for the sonocatalysis of various azo dyes. Meng and Oh (2011) prepared Fe-fullerene/TiO_2 catalyst for the sonocatalytic degradation of methylene blue. La^{3+} doping of TiO_2 particles synthesized by a sol–gel route was found to increase the catalytic efficiency in the sonocatalysis of amaranth (Song et al. 2011). Sonophotolytic advanced oxidation treatment has been found to be efficient in the removal of azo dye reactive black 5 (Zhou et al. 2011). However, organic ligands affected the performance.

3.2 Improving Disinfection

Studies have shown that high power ultrasound, operated at low frequencies is an effective means for the disintegration of bacterial cells (Blume and Neis 2004). However, disinfection by ultrasonication alone requires very high energy. Thus, generally it cannot be considered as an alternative to conventional disinfection for economical aspects. Therefore, ultrasonication should be used together with other techniques. For instance, the combination of a short ultrasonication and a subsequent ultraviolet treatment is cost-efficient (Blume and Neis 2004).

Ultrasonication combined with chlorination significantly improved the biocidal action. These results suggest that ultrasound could be used in conjunction with chemical treatments to achieve a reduction in the quantity of bactericide required for water treatment (Mason et al. 2003).

In another study by Joyce et al. (2003), 40 kHz ultrasound was used in conjunction with electrolysis to disinfect saline solution. The results show that sonication appears to amplify the effect of electrolysis: (i) ultrasound enhanced mixing of bacterial suspensions in the vicinity of the electrode surface where the hypochlorite is being generated; (ii) the mechanical action of cavitation damage and weakening the bacterial cell wall, thus making them more susceptible to an attack by hypochlorite; (iii) the cleaning action of ultrasound on the electrode surface preventing fouling build up, thus maintaining a more efficient electrolysis. Obviously, the combination of both treatments is significantly better than sonication or electrolysis alone.

3.3 Intensifying Electrocoagulation

The application of 22 kHz ultrasound to enhance electrocoagulation of effluents to remove surfactants was investigated by Sister and Kirshankova (2005). Theoretically, the main benefits that ultrasound brings during the electrocoagulation treatment are: (1) savings in electrical energy because the free radicals and outgassing effect on the cavitation raise the electrical conductivity, thus enabling the maintaining of a constant current at lower voltages, and (2) intensified electroflotation resulting from the gas formation. The combination with ultrasonication performed better effects than electrocoagulation alone. The optimum conditions found out are: the current density of $1–1.5$ A/dm^2, and time of joint ultrasonic and electrocoagulation treatment of $8–12$ min, with the ultrasonic power density of $1–2$ W/cm^2.

3.4 Enhancing Membrane Filtration

According to Kyllönen et al.'s review (2005), ultrasound irradiation can provide enhancement in the membrane filtration of wastewaters. It increases the flux primarily by breaking the cake layer at the membrane surface. Liquid jets produced by cavitation served as a basis for ultrasonic membrane cleaning. Lower ultrasound frequencies have higher cleaning efficiencies than higher frequencies. Intermittent ultrasound irradiation resulted in the same flux as obtained using continuous irradiation, but intermittent ultrasound consumed less energy and prolonged the lifetime of the membranes used. Therefore, it can be considered to be a cost effective method of membrane cleaning. Ultrasound assisted filtration is also less dependent on the feed properties.

4 Sludge Stabilization

Anaerobic digestion is the most commonly applied process for the stabilization of sewage sludge. The process is more beneficial among several sludge stabilization methods, because it has the ability to produce a net energy gain in the form of methane gas leading to cost-effectiveness. However, anaerobic digestion is a very slow process, and large fermenters are necessary. Enhanced performance of the anaerobic process could be achieved by finding a pre-treatment to accelerate the slow and rate-determining hydrolysis. Compared to other pre-treatment methods, ultrasonication exhibits a great potential of not being hazardous to environment and being economically competitive (Mao et al. 2004).

Sonix^{TM} is a new technology utilizing high-power concentrated ultrasound for conditioning sludges prior to further treatment (Hogan et al. 2004). The studies, which used ultrasound energy at frequencies above 20 kHz to create cavitation in the secondary sludge, have proved that the use of ultrasound to enhance anaerobic digestion can be achieved at full scale, resulting effectively in the thickened waste activated sludge (typically difficult to digest) after sonication. The technology presents benefits in terms of improved solids destruction, substantial increases in gas production and better residual solids dewatering.

Mao et al. (2004) have conducted ultrasound treatment of primary and secondary sludges to improve the quality of sludges for the anaerobic digestion. Experiment results indicate that a significant reduction in particle size and an increase in soluble organics could be achieved. Thus, ultrasonication could offer a feasible treatment method to efficiently disintegrate sludge. They also found that ultrasound treatment could be influenced by sonication density and solid concentration. The higher the sonication power employed, the more sludge particles are ruptured and more completely the structure is deteriorated. Also, a higher ultrasound density required less specific energy to derive a better sonication treatment.

Sludge disintegration was most significant at low frequencies. Low-frequency ultrasound created large cavitation bubbles, upon which collapse initiates powerful jet streams exerting strong shear forces in the liquid. Short sonication time resulted in sludge floc deagglomeration without the destruction of bacteria cells. Longer sonication brought about the breakup of cell walls. The sludge solids were disintegrated, and dissolved organic compounds were released (Yin et al. 2004).

5 Sediment and Soil Remediation

5.1 Heavy Metals Removal

Ultrasound has been used for a long time to the enhance precious metal recovery process by the cleaning action that removes an unwanted clay coating from raw ore, and accelerates leaching of minerals from the ore, as well as improves filtration rates (Newman et al. 1997). The collapse of cavitation bubbles near a solid can produce microjets which can cause the solid surface to pit and erode, promoting mechanical detachment. Furthermore, shock waves from cavitation in liquid–solid slurries can result in high-velocity inter-particle collisions which can also contribute to particle size reduction. In addition, the "cavities" or areas of low pressure in ultrasonic cavitation provide a sink of low concentration or partial pressure of the contaminant where adsorbed material will desorb (Meegoda and Perera 2001).

An investigation of ultrasonic treatment of polluted solid medium was carried out by Newman et al. (1997) more than 10 years ago. In that research, granular pieces of brick contaminated with copper oxide were used as a model for contaminated soil. Soil-washing was conducted by passing water across the substrate on an ultrasonically (20 kHz) shaken tray. This ultrasonic treatment considerably enhanced the process with 40% reduction in copper content, compared to only 6% reduction by conventional shaking (Newman et al. 1997).

Meegoda and Perera (2001) studied the 20 kHz sonication coupled with extraction using vacuum pressure in an integrated multi-step technology, to remove heavy metal contaminants from dredged residues. Ultrasound power, soil-to-water ratio, dwell time and vacuum pressure were considered to be important factors that influence the treatment process. The study showed that the proposed treatment technique is effective and economical for sediments with lower clay contents (only the silt fraction had a considerable metal removal while the clay fraction was insensitive to the treatment). A maximum removal of 83% was obtained for silt fraction at 1200 W power, 1:50 soil-to-water ratio and 90 min of dwell time (Meegoda and Perera 2001).

Kyllönen et al. (2004) researched power ultrasound as an aiding method for the mineral processing technique, which has recently been developed for the remediation of soil contaminated by heavy metals. Power ultrasound was used to disperse the soil to remove metals and metal compounds from soil particle surfaces instead of attrition conditioning. The soil diluted with water was treated using 22 kHz ultrasound power of 100 W up to 500 W. The effect of different ultrasonic treatment time and pulsation of ultrasound were studied on the purity of sink and float fractions in heavy medium separation process, screen fractions, and mineral concentrates and tailings from flotation process. Ultrasound enhanced the remediation of soil fractions in all the studied cases (Kyllönen et al. 2004).

5.2 Organic Decontamination

Although ultrasonic applications in environmental areas are still in lab-scale and developing stage, they are increasing rapidly and attracting more and more interest because of many advantages they offer: environmental friedliness (no toxic chemicals are used or produced), low energy demand, and compact and transportable method that can be used on-site. Environmental remediation by ultrasonication is involved mostly in organic pollutant destruction, through thermal decomposition (pyrolysis) and the formation of oxidative species like hydroxyl radical that enhance the mineralization of pollutants. Moreover, in soil treatment, ultrasonic waves increase the porosity of the soil and percolation rate, thus accelerating the desorption and facilitating the removal of entrapped contaminants. On the other hand, ultrasound applied in environmental analysis also provides benefits, such as shorter time, simplified procedure, and higher purity of the final product.

5.2.1 Effect of Ultrasound on Desorption of Organic Contaminants

Similar to the application of ultrasonic leaching for metal removal, ultrasound has been known for promoting organic desorption from soils and sediments (Pee 2008; Kim and Wang 2003). According to Feng and Aldrich (2000), the likely mechanism of ultrasonic desorption can be explained by considering the different effects of ultrasound in heterogenous media. Firstly, the high temperatures in localized hot spots enhance the breaking of physical bonds between the adsorbate (contaminants) and the adsorbent surface. Secondly, acoustic cavitation produces high-speed microjets and high-pressure shock waves that impinge on the surface and erode the adsorbate (Suslick et al. 1987; Stephanis et al. 1997). Finally, ultrasound produces acoustic vortex microstreaming within the pores of the solid particles and the solid–liquid interface. This phenomenon arises by the increase in the momentum brought about as the liquid absorbs energy from the propagating sound waves, even in the absence of cavitation (Ley and Low 1989). These effects may possibly be the cause of enhanced desorption rates (Feng and Aldrich 2000).

In their study, Feng and Aldrich (2000) investigated the influence of factors such as slurry concentration, ultrasonic power intensity, duration of irradiation, particle size, diesel content, slurry pH, salinity, and surfactant dosage during the remediation of stimulated soil contaminated with diesel fuel in the presence of ultrasound. Ultrasonic treatment performed more effectively than high-speed mechanical agitation at the same energy input. However, prolonged ultrasonic irradiation could not increase the efficiency of diesel removal, possibly due to the equilibrium between the desorption and re-adsorption of these hydrocarbons from and onto the particle surfaces. A multistage sonochemical treatment process was proposed for the remediation of sand contaminated with diesel, as better results were obtained from this approach than from the single-stage treatment process.

5.2.2 Effect of Ultrasound in Destroying Organic Contaminants

Ultrasonication not only assists the desorption of the contaminants from the soil, but also promotes the formation of the strong oxidant, ·OH radical (Flores et al. 2007). Ultrasonic energy can destroy the contaminants through oxidation by free radicals and pyrolysis processes, not only transport the contaminants from one place to another like in conventional soil washing.

Collings et al. (2006) have developed high power ultrasound to destroy persistent organic pollutants (POPs) in soils and sediments. They have worked successfully on major contaminants, atrazine, simazine, total petroleum hydrocarbons, DDT, lindane, endosulfan, 2,4,5-T, tetrachloronaphthalene and TBT. The range of contaminants in their study is sufficiently broad to suggest that high power ultrasound will be effective for most adsorbed large molecules. The results indicate several advantages of high power ultrasonic technology compared with conventional methods. These include high destruction rates, the lack of dangerous breakdown products, and low energy demand leading to low-cost. Moreover, the technology can be made quite compact and transportable, allowing on-site treatment.

The feasibility of ultrasonication on the treatment of different kinds of highly contaminated soils (synthetic clay, natural farm clay, kaolin) (Shrestha et al. 2009) was investigated by using two target POPs; hexachlorobenzene (HCB) and phenanthrene (PHE). Experimental results showed that ultrasonication has a potential to reduce the high concentrations of these POPs. The treatment of soil by ultrasonication requires some amount of water for sonochemistry effects to perform. The reasonable moisture ratio of the slurry could be from 2:1 to 3:1 water and soil, the higher the better. Particularly kaolin needed higher amount of water than other clays to perform well. The removal efficiency increased slightly after a long ultrasonication time. Ultrasonication did not affect the pH values of slurries. The heating and irritating noise problems of ultrasonication should be considered carefully in larger scale applications. The removal rates of POPs in soils vary with soil type, power and the frequency of the ultrasound applied.

5.3 Ultrasonication as Assistant Process in Organic Contaminated Soil Remediation

In most of the cases, ultrasound is used as a supplemental method to enhance the soil remediation process.

5.3.1 Ultrasonically Enhanced Soil Flushing

An ultrasonically enhanced soil-flushing method for in situ remediation of the ground contaminated by non-aqueous phase liquid (NAPL) hydrocarbons was investigated by Kim and Wang (2003). Crisco Vegetable Oil was chosen as the model compound. The soil-flushing tests were conducted in two conditions—without ultrasound and with 20 kHz ultrasonic waves. Experimental results indicated that ultrasonication can enhance oil removal considerably. The degree of enhancement depends on factors such as ultrasonic power, water washing flow rate, and soil type. Increasing ultrasonic power will increase pollutant extraction only up to the level where cavitation occurs. The effectiveness of ultrasonication decreases with flushing rate but eventually becomes constant under higher flow rates (Kim and Wang 2003).

Mason et al. (2004) reported some laboratory research on ultrasonic soil washing of organic contaminants like pesticide DDT, PCB and polycyclic aromatic hydrocarbon (PAH). Initial concentrations of DDT (250 ppm), PCB (250 ppm) and PAH (400 ppm) in sand (200 g contaminated fine sand in 200 g water) were removed ultrasonically (20 kHz, 170 W) by 70% after 10, 25 and 3 min, respectively. The potential for the scale-up of the soil washing using acoustic energy was also reported. Two basic mechanisms for acoustically enhanced soil washing have been suggested: the abrasion of surface cleaning and the leaching out of more deeply entrenched material. According to Mason, the factors contributing towards improvements in the efficiency under the influence of ultrasound include: (i) the high-speed microjet formation during the asymmetric cavitation bubble collapse in the vicinity of the solid surface, enhancing transport rates and also increasing the surface area through surface pitting; (ii) particle fragmentation through collisions increasing the surface area; and (iii) the enhancement of diffusion by the ultrasonic capillary effect.

Shrestha et al. (2009) showed the feasibility of ultrasound on the treatment of contaminated soils (synthetic clay, natural farm clay, and kaolin) by using two target POPs: HCB and PHE. The soils were highly contaminated (500 mg/kg). The reasonable moisture ratio of the slurry could be in the range 2:1–3:1. The great advantage of this process was that there was no change in the pH values of the soils.

5.3.2 Ultrasonically Assisted Advanced Oxidative Soil Remediation

A new process for the remediation of soil contaminated with organic compounds (toluene and xylenes) was proposed by Flores et al. (2007). The innovation combined the advanced oxidation method using Fenton-type catalyst with the application of ultrasonic energy (47 kHz, 147 W, 10 min duration time for 20 g soil in 40 g aqueous solution). Experimental results showed that the application of ultrasound not only assists the desorption of the contaminants from the soil, but also promotes the formation of hydroxyl radicals, which are the main oxidant agent involved in the decontamination process. The global efficiency of the process was noticeable enhanced when applying ultrasonic energy, due to a synergistic effect in conjunction with the hydrogen peroxide concentration and Fenton catalyst (Flores et al. 2007).

5.3.3 Ultrasonically Enhanced Electrokinetic Remediation

Previous studies showed that electrokinetic technique was applied to remove mainly heavy metals, and the ultrasonic technique was applied to remove mainly organic substances in the contaminated soil. Thus, the combination of the two techniques can predictably be helpful. Chung and Kamon (2005) studied electrokinetic and ultrasonic remediation technologies for the removal of heavy metal and PAH in contaminated soils. The study emphasized the coupled effects of the electrokinetic and ultrasonic techniques on the migration, as well as the clean-up of contaminants in soils. Natural clay was used as a test specimen, while Pb and phenanthrene were used as contaminants. Pb is a positively charged ionic contaminant; on the other hand, phenanthrene is a neutrally charged nonionic contaminant. The ultrasonic processor had a maximum power output of 200 W with a frequency of excitation equal to 30 kHz.

When ultrasonic energy was applied into contaminated soil, the viscosity of fluid phase decreased and flow rate increased, the molecular movement increased, sorbed contaminants mobilized, the cavitation developed, and the porosity and permeability increased. The removal efficiency of the contaminant was higher for the combined electrokinetic–ultrasonic test than for the simple electrokinetic test alone. Therefore, the introduction of an enhancement technique, like an ultrasonic process, into an electrokinetic process could be effective for increasing the contaminant removal rate of the contaminated soil.

Tests were also conducted using ultrasound alone, using ultrasound as an enhancement for the electrokinetic test, and using electrokinetic test alone to compare the removal performance of the three persistent organic pollutants, hexachlorobenzene, phenanthrene, fluoranthene (Pham et al. 2009a, b) and chrysene from low permeability kaolin in reactors and pans, with and without iron anodes (Shrestha et al. 2010). The results of these experiments show that the combined electrokinetic and ultrasonic treatment did prove a more positive coupling effect in PAHs removal than each single process alone, though the level of enhancement was not significant. Results indicated that the removal was more effective with lower concentrations of organic pollutants. The average removal was better in the pan experiment with EKUS with iron anode (Shrestha et al. 2010). This might be due to the increase in the electroconductivity by iron ions. The assistance of ultrasound in the electrokinetic remediation can help to reduce POPs from clayey soil by improving the mobility of hydrophobic organic compounds and degrading these contaminants through pyrolysis and oxidation. Ultrasonication also sustains higher current and increases electroosmotic flow in the combined EK-US test than in the EK test alone.

5.3.4 Ultrasonically Enhanced Activated Carbon Amendment

The addition of carbon particles to sediment provides strong sorption sites for the hydrophobic organic contaminants and reduces these freely dissolved compounds' concentrations. Thus, a powdered activated carbon (PAC) amendment assisted with sonication was used to reduce the bioaccessibility of polycyclic aromatic hydrocarbons in three creosote contaminated sediments (Pee 2008). The study revealed that the sonochemically induced switching of phenanthrene and pyrene from sediment to PAC was more effective than the mechanical mixing in decreasing the bioavailability of these PAHs. The enhancement effect performed in sediment treated with sonication was explained to be attributed to the facilitation of desorption of PAHs through localized turbulent liquid movement, microjets formation and particles fragmentation.

5.3.5 Ultrasonically Enhanced Surfactant Aided Soil Washing

The use of ultrasound as an enhancement mechanism in the surfactant-aided soil-washing process was examined by conducting desorption tests of soils contaminated with naphthalene or diesel-oil (Seungmin et al. 2007). The experiments were conducted to elucidate the effect of ultrasound on the mass transfer from soil to the aqueous phase using naphthalene-contaminated soil. In addition, the use of ultrasound for the diesel-oil-contaminated soil was investigated under a range of conditions of surfactant concentration, sonication power, duration, soil/liquid ratio, particle size, and initial diesel-oil concentration. The ultrasound used in the soil-washing process significantly enhanced the mass transfer rate from the solid phase to the aqueous phase. The removal efficiency of diesel-oil from the soil phase generally increased with longer sonication time, higher power intensity, and large particle size.

6 Air Pollution Control

The application of ultrasound in air pollution control is based on an acoustic agglomeration phenomenon that makes small particles precipitated for easy removal. Acoustic agglomeration is a process in which high intensity sound waves produce relative motion and collisions among fine particles suspended in gaseous media. In an acoustic field, fine particles suspended in the air will migrate to the nodes of the sound wave, becoming concentrated. Once the particles collide, they tend to adhere together to form larger agglomerated particles (Mason 2007a, b). More details on acoustic agglomeration mechanisms can be found in Hoffmann's article (2000). In general, acoustic agglomeration can be conducted through two approaches; with low frequency and high frequency (ultrasound) sonication. While low frequency acoustic field is more cost and energy efficient, high frequency acoustic (ultrasonic) agglomeration might achieve better particle retention efficiency, especially for very small particles in submicron range (Hoffmann 2000).

Recently, Riera et al. (2003) studied the effect of humidity on the acoustic agglomeration of submicron particles in diesel exhausts at 21 kHz ultrasound. The presence of humidity raised the agglomeration rate by decreasing the number particle concentration up to 56%. The results confirmed the benefit of using high-power ultrasound together with humidity to enhance the agglomeration of particles much smaller than 1 μm (Riera et al. 2003).

7 Environmental Analysis

The use of ultrasound in environmental analysis brings many benefits such as shorter analysis time, simplified manipulation and higher purity of the final product (Chemat et al. 2004).

7.1 Assisting Microwave Digestion

Conventional digestion is often carried out using a prolonged heating and stirring in strong acid solution, taking at least several hours. In recent decades, microwave heating has been used in analytical and organic laboratory practices as a very effective and non-polluting method of activation. Moreover, nowadays, the simultaneous microwave and ultrasound irradiation has been recognized as a new technique for atmospheric pressure digestion of solid and liquid samples in chemical analysis. The coupling microwave-ultrasound gives significant improvements such as the reduction of digestion time, the reduction of the quantity of reagents, and the reduction of contamination. In addition, the process can be totally automatic and safer. The combination of these two types of irradiation in physical processes like digestion, dissolution and extraction appears very promising. (Chemat et al. 2004).

7.2 Assisting Solvent Extraction

Mecozzi et al. (2002) proposed an accelerated ultrasound assisted procedure for the extraction of the available humic substance from marine sediments. The main advantage of the ultrasonic method is the shorter time of extraction, taking only 30 min in contrast to the 24 h required by the shaking method. In addition, the extracts obtained were homogenous and qualitative.

Ultrasonic solvent extraction of the organochlorine pesticide (OCP) including DDT, DDE, Dieldrin, methoxychlor, and mirex from soil was developed by Tor et al. (2006). The results obtained indicated that the ultrasonic solvent extraction method could be efficiently applied in order to extract OCP from soils with the recovery rate up to over 92%. Moreover, the ultrasonic solvent extraction is more rapid, as time consumption was reduced approximately 75 and 82% compared to conventional shake-flash and soxhlet extraction. The solvent consumption is also significantly lower than in soxhlet extraction, with 67% reduction.

Another new sample pretreatment technique, ultrasound-assisted headspace liquid-phase microextraction was successfully applied by Xu et al. (2007) to determine chlorophenols in real aqueous samples. With good recoveries ranging up to 100%, the proposed method demonstrated very promisingly suitable for the analysis of trace volatile and semivolatile pollutants. Moreover, its advantages over the conventional headspace liquid-phase microextraction include simple setup, ease of operation, rapidness, sensitivity, precision, and no cross-contamination (Xu et al. 2007).

8 Conclusion

Sonochemistry has been rapidly developing in recent years. Its potential in environmental applications is drawing more and more attention. Ultrasonic bath is widely used in analytical laboratories as an efficient method for solubilization, extraction assistance, and cleaning. Moreover, the utilization of ultrasound in environmental protection covers a broad range of applications: water treatment, soil remediation, and air cleaning. Among these environmental remediations, organic water decontamination is perhaps the most extensively researched, due to the fact that chemically ultrasonic effects work best in an aqueous medium because of the free radical formation during water sonolysis. On the other hand, the physical effects of ultrasound are recognized in membrane filtration, sediment heavy metal removal, dewatering, and air cleaning. However, it is generally accepted that ultrasonication alone cannot be a very cost-efficient technique. Ultrasound should rather be combined with other specific methods or work as an assistant for enhanced performance. Moreover, other physical impacts like heating, noise during ultrasonic process, and economic factors should also be considered, especially in practical scale-up systems. In general, as a part of a young and interesting science, the applications of ultrasound in the environmental and other green technology have a promising future.

Acknowledgments Sari Silventoinen is thanked for language revision and Jurate Virkutyte for useful comments.

References

Abramov VO, Abramov OV, Gekhman AE, Kuznetsov VM, Price GJ (2006) Ultrasonic intensification of ozone and electrochemical destruction of 1, 3-dinitrobenzene and 2, 4-dinitrotoluene. Ultrason Sonochem 13:303–307

Abu-Hassan MA, Kim JK, Metcalfe IS, Mantzavinos (2006) Kinetics of low frequency sonodegradation of linear alkylbenzene sulfonate solutions. Chemosphere 62:749–755

Adewuyi YG (2001) Sonochemistry: environmental science and engineering applications. Ind Eng Chem Res 40:4681–4715

Belgiorno V, Rizzo L, Fatta D, Rocca CD, Lofrano G, Nikolaou A, Naddeo V, Meric S (2007) Review on endocrine disrupting-emerging compounds in urban wastewater: occurrence and removal by photocatalysis and ultrasonic irradiation for wastewater reuse. Desalination 215:166–176

Blume T, Neis U (2004) Improved wastewater disinfection by ultrasonic pre-treatment. Ultrason Sonochem 11:333–336

Casadonte DJ, Flores M, Petrier C (2005) Enhancing sonochemical activity in aqueous media using power-modulated pulsed ultrasound: an initial study. Ultrason Sonochem 12:147–152

Chemat S, Lagha A, Amar HA, Chemat F (2004) Ultrasound assisted microwave digestion. Ultrason Sonochem 11:5–8

Chung HI, Kamon M (2005) Ultrasonically enhanced electrokinetic remediation for removal of Pb and phenanthrene in contaminated soils. Eng Geol 77:233–242

Collings AF, Farmer AD, Gwan PB, Sosa Pintos AP, Leo CJ (2006) Processing contaminated soils and sediments by high power ultrasound. Miner Eng 19:450–453

Dewulf J, Langenhove HV, Visscher AD, Sabbe S (2001) Ultrasonic degradation of trichloroethylene and chlorobenzene at micromolar concentration: kinetics and modeling. Ultrason Sonochem 8:143–150

Drijvers D, Baets RD, Visscher AD, Langenhove HV (1996) Sonolysis of trichloroethylene in aqueous solution: volatile organic intermediates. Ultrason Sonochem 3:83–90

Emery RJ, Papadaki M, Mantzavinor D (2003) Sonochemical degradation of phenolic pollutants in aqueous solutions. Environ Technol 24:1491–1500

Entezari MH, Petrier C, Devidal P (2003) Sonochemical degradation of phenol in water: a comparison of classical equipment with a new cylindrical reactor. Ultrason Sonochem 10:103–108

Feng D, Aldrich C (2000) Sonochemical treatment of simulated soil contaminated with diesel. Adv Environ Res 4:103–112

Flores R, Blass G, Dominguez V (2007) Soil remediation by an advanced oxidative method assisted with ultrasonic energy. J Hazard Mater 140:399–402

Gao J, Jiang R, Wang J, Kang P, Wang B, Li Y, Li K, Zhang X (2011) The investigation of sonocatlytic activity of Er^{3+}:$YAlO_3$/TiO_2-ZnO composite in azo dyes degradation. Ultrason Sonochem 18:541–548

Gogate PR (2008) Treatment of wastewater streams containing phenolic compounds using hybrid techniques based on cavitation: a review of the current status and the way forward. Ultrason Sonochem 15:1–15

Goskonda S, Catallo WJ, Junk T (2002) Sonochemical degradation of aromatic organic pollutants. Waste Manag 22:351–356

He Z, Song S, Ying H, Xu L, Chen J (2007) p-Aminophenol degradation by ozonation combined with sonolysis: Operating conditions influence and mechanism. Ultrason Sonochem 14:568–574

Hoffmann TL (2000) Environmental implications of acoustic aerosol agglomeration. Ultrasonics 38:353–357

Hoffmann MR, Hua I, Höchemer R (1996) Application of ultrasonic irradiation for the degradation of chemical contaminants in water. Ultrason Sonochem 3:163–172

Hogan F, Mormede S, Clark P, Crane M (2004) Ultrasonic sludge treatment for enhanced anaerobic digestion. Water Sci Technol 50:25–32

Hua I, Pfalzer-Thompson U (2001) Ultrasonic irradiation of carbofuran: decomposition kinetics reactor characterization. Wat Res 35:1445–1452

Ince NH, Tezcanli G, Belen RK, Apikyan G (2001) Ultrasound as a catalyzer of aqueous reaction systems: the state of the art and environmental applications. Appl Catal B 29:167–176

Jiang Y, Petrier C, Waite TD (2002a) Effect of pH on the ultrasonic degradation of ionic aromatic compounds in aqueous solution. Ultrason Sonochem 9:163–168

Jiang Y, Petrier C, Waite TD (2002b) Kinetics and mechanisms of ultrasonic degradation of volatile chlorinated aromatics in aqueous solutions. Ultrason Sonochem 9:317–323

Joseph JM, Destaillats H, Hung H, Hoffmann MR (2000) The sonochemical degradation of azobenzene and related azo dyes: rate enhancement via Fenton's reactions. J Phys Chem A 104:301–307

Joyce E, Mason TJ, Phull SS, Lorimer JP (2003) The development and evaluation of electrolysis in conjunction with power ultrasound for the disinfection of bacterial suspension. Ultrason Sonochem 10:231–234

Kidak R, Ince NH (2006) Ultrasonic destruction of phenol and substituted phenols: a review of current research. Ultrason Sonochem 13:195–199

Kim YU, Wang MC (2003) Effect of ultrasound on oil removal from soils. Ultrasonics 41:539–542

Kyllönen H, Pirkonen P, Hintikka V, Parvinen P, Grönroos A, Sekki H (2004) Ultrasonically aided mineral processing technique for remediation of soil contaminated by heavy metals. Ultrason Sonochem 11:211–216

Kyllönen HM, Pirkonen P, Nyström M (2005) Membrane filtration enhanced by ultrasound—a review. Desalination 181:319–335

Ley SV, Low CMR (1989) Ultrasound in Synthesis, Chap. 2. Springer-Verlag, Berlin

Manariotis ID, Karapanagioti HK, Chrysikopoulos CV (2011) Degradation of PAHs by high frequency ultrasound. Wat Res 45:2587–2594

Mao T, Hong SY, Show KY, Tay JH, Lee DJ (2004) A comparison of ultrasound treatment on primary and secondary sludges. Water Sci Technol 50:91–97

Mason TJ (2007a) Review—developments in ultrasound—non-medical. Prog Biophys Mol Biol 93:166–175

Mason TJ (2007b) Sonochemistry and the environment—providing a "green" link between chemistry, physics and engineering. Ultrason Sonochem 14:476–483

Mason TJ, Joyce E, Phull SS, Lorimer JP (2003) Potential uses of ultrasound in the biological decontamination of water. Ultrason Sonochem 10:319–323

Mason TJ, Collings A, Sumel A (2004) Sonic and ultrasonic removal of chemical contaminants from soil in the laboratory and on a large scale. Ultrason Sonochem 11:205–210

Matouq MA, Al-Anber ZA (2007) The application of high frequency ultrasound waves to remove ammonia from simulated industrial wastewater. Ultrason Sonochem 14:393–397

Mecozzi M, Amici M, Pietrantonio E, Romanelli G (2002) An ultrasound assisted extraction of the available humic substance from marine sediments. Ultrason Sonochem 9:11–18

Meegoda JN, Perera R (2001) Ultrasound to decontaminate heavy metals in dredged sediments. J Hazard Mater 85:73–89

Meng Z-D, Oh W-C (2011) Sonocatalytic degradation and catalytic activities for MB solution of Fe treated fullerene/TiO_2 composite with different ultrasonic intensity. Ultrason Sonochem 18(2011):757–764

Naddeo V, Belgiorno V, Kassinos D, Mantzavinos D, Meric S (2010) Ultrasonic degradation, mineralization and detoxification of diclofenac in water: optimization of operating parameters. Ultrason Sonochem 17:179–185

Nakui H, Okitsu K, Maeda Y, Nishimura R (2007) Hydrazine degradation by ultrasonic irradiation. J Hazard Mater 146:636–639

Newman AP, Lorimer JP, Mason TJ, Hutt KR (1997) An investigation into the ultrasonic treatment of polluted solids. Ultrason Sonochem 4:153–156

Papadaki M, Emery RJ, Abu-Hassan MA, Diaz-Bustos A, Metcalfe Mantzavinos D (2004) Sonocatalytic oxidation processes for the removal of contaminants containing aromatic rings from aqueous effluents. Sep Sci Technol 34:35–42

Pee GY (2008) Sonochemical remediation of freshwater sediments contaminated with polycyclic aromatic hydrocarbons. PhD dissertation, The Ohio State University

Pham TD, Shrestha RA, Sillanpää M (2009a) Electrokinetic and ultrasonic treatment of kaoline contaminated. POPs Sep Sci Technol 44(10):2410–2420

Pham TD, Shrestha RA, Virkutyte J, Sillanpää M (2009b) Recent studies in environmental applications of ultrasound. J Environ Eng Sci 36:1849–1858

Riera-Franco de Sarabia E, Elvira-Segura L, Gonzalez-Gomez I, Rodriguez-Maroto JJ, Munoz-Bueno R, Dorronsoro-Areal JL (2003) Investigation of the influence of humidity on the ultrasonic agglomeration of submicron particles in diesel exhausts. Ultrasonics 41:277–281

Sáez V, Esclapez MD, Bonete P, Walton DJ, Rehorek A, Louisnard O, González-García J (2011a) Sonochemical degradation of perchloroethylene: the influence of ultrasonic variables, and the identification of products. Ultrason Sonochem 18:104–113

Sáez V, Tudela I, Esclapez MD, Bonete P, Louisnard O, González-García J (2011b) Sonochemical degradation of perchloroethylene: the influence of ultrasonic variables, and the identification of products. Chem Eng J 168:649–655

Sangave PC, Pandit AB (2004) Ultrasound pre-treatment for enhanced biodegradability of the distillery wastewater. Ultrason Sonochem 11:197–203

Seungmin N, Young-Uk K, Jeehyeong K (2007) Physiochemical properties of digested sewage sludge with ultrasonic treatment. Ultrason Sonochem 14:281–285

Shrestha RA, Pham TD, Sillanpää M (2009) Effect of ultrasound on removal of persistent organic pollutants (POPs) from different types of soils. J Hazard Mater 170:871–875

Shrestha RA, Pham TD, Sillanpää M (2010) Electro ultrasonic remediation of polycyclic aromatic hydrocarbons from contaminated soil. J Appl Electrochem 40(7):1407–1413

Singla R, Grieser F, Ashokkumar M (2011) The mechanism of sonochemical degradation of a cationic surfactant in aqueous solution. Ultrason Sonochem 18:484–488

Sister VG, Kirshankova EV (2005) Ultrasonic techniques in removing surfactants from effluents by electrocoagulation. Chem Petrol Eng 41:553–556

Song L, Chen C and Zhang S (2011) Sonocatalytic degradation of amaranth catalyzed by La^{3+} doped TiO_2 under ultrasonic irradiation. Manuscript. Sonochemistry Research , The Sonochemistry Centre at Coventry University. http://www.sonochemistry.info/research.html. Accessed Oct 2007

Stephanis CG, Hatiris JG, Mourmouras DE (1997) The process (mechanism) of erosion of soluble brittle materials caused by cavitation. Ultrason Sonochem 4:269–271

Suslick KS (2006) Summary of sonochemistry and sonoluminescence. Suslick Research Group University of Illinois. http://www.scs.uiuc.edu/suslick/execsummsono.html/. Accessed Mar 2006

Suslick KS, Casadonte DJ, Green MLH, Thompson ME (1987) Effects of high intensity ultrasound on inorganic solids. Ultrasonics 25:56–61

Teo KC, Xu Y, Yang C (2001) Sonochemical degradation for toxic halogenated organic compounds. Ultrason Sonochem 8:241–246

Tezcanli-Guyer G, Ince NH (2003) Degradation and toxicity reduction of textile dyestuff by ultrasound. Ultrason Sonochem 10:235–240

Tor A, Aydin ME, Özcan S (2006) Ultrasonic solvent extraction of organochlorine pesticides from soil. Anal Chim Acta 559:173–180

Vajnhandl S, Majcen Le Marechal A (2005) Review—ultrasound in textile dyeing and the decolouration/mineralization of textile dyes. Dyes Pigment 65:89–101

Visscher AD, Langenhove HV, Eenoo PV (1997) Sonochemical degradation of ethylbenzene in aqueous solution: a product study. Ultrason Sonochem 4:145–151

Wang YQ, Pan L, Tao J, Wang T (2011) Bioactive porous titania formed by two-step chemical treatment of titanium substrates under high intensity ultrasonic field. Surf Eng 27:46–50

Wayment DG, Casadonte DJ (2002) Frequency effect on the sonochemical remediation of alachlor. Ultrason Sonochem 9:251–257

Xu H, Liao Y, Yao J (2007) Development of a novel ultrasound-assisted headspace liquid-phase microextraction and its application to the analysis of chlorophenols in real aqueous samples. J Chromatogr A 1167:1–8

Yao JJ, Gao NY, Deng Y, Ma Y, Li HJ, Xu B, Li L (2010) Sonolytic degradation of parathion and the formation of byproducts. Ultrason Sonochem 17:802–809

Yasman Y, Bulatov V, Gridin VV, Agur S, Galil N, Armon R, Schechter I (2004) A new sono-electrochemical method for enhanced detoxification of hydrophilic chloroorganic pollutants in water. Ultrason Sonochem 11:365–372

Yasman Y, Bulatov V, Rabin I, Binetti M, Schechter I (2006) Enhanced electro-catalytic degradation of chloroorganic compounds in the presence of ultrasound. Ultrason Sonochem 13:271–277

Yin X, Han P, Lu X, Wang Y (2004) A review on the dewaterability of bio-sludge and ultrasound pretreatment. Ultrason Sonochem 11:337–348

Zhou T, Lim T-T, Wu X (2011) Sonophotolytic degradation of azo dye reactive *black* 5 in an ultrasound/UV/ferric system and the roles of different organic ligands. Wat Res 45:2915–2924. doi: 10.1016/j.watres.2011.03.008